D1221537

PowerKids Readers:

EARTH MOVERS ™

Backhoes

Joanne Randolph

The Rosen Publishing Group's
PowerKids Press ™
New York

1

For Ryan, with love

Published in 2002 by The Rosen Publishing Group, Inc.
29 East 21st Street, New York, NY 10010

Copyright © 2002 by The Rosen Publishing Group, Inc.

First Edition

Book Design: Michael Donnellan

Photo Credits: p. 5 © CORBIS/Carl Corey; p. 7 © SuperStock; pp. 9, 15 © Highway Images/Genat; p. 11 © CORBIS/Raymond Gehman; pp. 13, 19 © CORBIS; p. 17 © CORBIS/Chinch Gryniewicz; p. 21 © CORBIS/Kevin R. Morris.

Randolph, Joanne.
Backhoes / Joanne Randolph—1st ed.
    p. cm—(Earth movers)
Includes bibliographical references and index.
ISBN 0-8239-6029-3
1. Backhoes—Juvenile literature. [1. Backhoes.] I. Title.
TA735 .R36 2002
629.225-dc21

00-013008

Manufactured in the United States of America

Contents

This is a backhoe.

A backhoe uses two tools to do its job. One tool is in the front and one is in the back.

The front of a backhoe
has a bucket. The bucket
is used for pushing
and lifting.

9

The back of a backhoe
has a special shovel.

The shovel has teeth that help dig up the ground.

This backhoe is helping
to build a road.

15

This backhoe is helping
the men dig a hole
for pipes.

Special tools can be attached to a backhoe. This backhoe has a claw attached to it. The claw picks up garbage from the ground.

Backhoes are very
useful trucks.

Words to Know

backhoe

bucket

claw

shovel

Here are more books to read about backhoes:

Diggers and Other Construction Machines (Cutaway Series)
By Jon Richards
Copper Beach Books

Diggers and Dump Trucks (Eye Openers)
By Angela Royston
Little Simon

To learn more about backhoes, check out this Web site:
www.howstuffworks.com/hydraulic.htm

Index

Word Count: 108
Note to Librarians, Teachers, and Parents

PowerKids Readers are specially designed to help emergent and beginning readers build their skills in reading for information. Simple vocabulary and concepts are paired with photographs of real kids in real-life situations or stunning, detailed images from the natural world around them. Readers will respond to written language by linking meaning with their own everyday experiences and observations. Sentences are short and simple, employing a basic vocabulary of sight words, as well as new words that describe objects or processes that take place in the natural world. Large type, clean design, and photographs corresponding directly to the text all help children to decipher meaning. Features such as a contents page, picture glossary, and index help children get the most out of PowerKids Readers. They also introduce children to the basic elements of a book, which they will encounter in their future reading experiences. Lists of related books and Web sites encourage kids to explore other sources and to continue the process of learning.